THE BIG iDEA

CRICK, WATSON,
AND
DNA

PAUL STRATHERN

ANCHOR BOOKS
DOUBLEDAY
New York London Toronto Sydney Auckland

AN ANCHOR BOOK
PUBLISHED BY DOUBLEDAY
a division of Random House, Inc.
1540 Broadway, New York, New York 10036

ANCHOR BOOKS, DOUBLEDAY, and the portrayal of an anchor are
trademarks of Doubleday, a division of Random House, Inc.

Crick, Watson, and DNA was originally published in the United Kingdom
by Arrow Books, a division of Random House U.K. Ltd. The Anchor
Books edition is published by arrangement with Arrow Books.

Library of Congress Cataloging-in-Publication Data

Strathern, Paul, 1940–
Crick, Watson, and DNA / Paul Strathern.
p.   cm. — (The Big idea)
Includes bibliographical references.
1. Genetics—History.   2. DNA—Research—History.   3. Genetics
Popular works.   4. Crick, Francis, 1916–   .   5. Watson, James D.,
1928–   .   I. Title.   II. Title: Crick, Watson, and DNA.
III. Series: Strathern, Paul, 1940–   Big idea.
QH428.S77   1999
576.5′09—dc21   99-26186
CIP

ISBN 0-385-49245-6

1   3   5   7   9   10   8   6   4   2

Contents

Introduction

THE GREAT scientific advance of the first half of the twentieth century was nuclear physics. Relativity and quantum theory began unlocking the secrets of the atom, discovering the ultimate matter of the universe. Nuclear physics became the cutting edge of human knowledge.

The mid-century discovery of the structure of DNA created an entirely new science. This was molecular biology, which began unlocking the secrets of life itself. Molecular biology now became the nuclear physics of the second half of the twentieth century.

The discoveries being made in this field (and the possible discoveries yet to be made) are transforming our entire conception of life. Like children, we have discovered the ultimate building blocks of life, and we are also learning how they can be taken apart. Once again, science has outstripped morality. We are acquiring dangerous knowledge, without any clear idea of how we should use it. As yet, we are barely grappling with the moral problems posed by nuclear physics (which may yet destroy us). Molecular biology is showing us how to transform life into almost anything.

Such scary possibilities were barely glimpsed by those who sought to discover "the secret of life." For them, this was one of the great scientific adventures. This adventure may have been pure in its aims, but those who took part in it were not immune from human frailty. All human life is here: ambition, supreme intelligence, folly, wishful thinking, incompetence, and sheer luck (both good and bad)—all had their part to play. The search for the secret of life proved no different from life itself. And the answer, when it was finally discovered, fell into the same category. The structure of DNA is fiendishly complex, astonishingly beautiful, and contains the seeds of tragedy.

Crick, Watson, and DNA

On the Way to DNA: A History of Genetics

UNTIL LITTLE OVER a century ago, genetics was mostly old wives' tales. People saw what happened, but had no idea how or why it happened.

References to genetics go back as far as biblical times. According to Genesis, Jacob had a method for making sure that his sheep and goats gave birth to spotted and speckled offspring. He did this by making them breed in front of sticks with strips of peeled bark which had a similar mottled effect.

More realistically, the Babylonians understood that for a date palm to be fruitful, pollen from the male palm had to be introduced to the pistils of the female palm.

The ancient Greek philosophers were the first to look at the world in a recognizably scientific fashion. As a result they produced theories about almost everything, and genetics was no exception. Aristotle's observations led him to conclude that the male and female do not make equal contributions to their offspring. Their contributions are qualitatively different: the female gives "matter," the male gives "motion."

A prevalent belief in ancient times held that if a female had previously mated and had progeny, the characteristics of their father would appear in the woman's subsequent progeny by any other male. This fairy story was even dignified with a pseudo-scientific name by the ancient Greeks, who called it telegony (meaning "distant-begetting").

A more interesting theory was pangenesis, which held that each organ and substance of the body secreted its own particles, which then combined to form the embryo.

Such beliefs recur in genetic theory through the centuries, in a manner curiously similar to the actual recurrence of genetic traits. (Pangenesis was to pop up for well over 2000 years, and was even accepted by Darwin.)

Biology, and with it genetics, crossed the threshold into science in the seventeenth century. This was almost entirely due to the microscope, which was invented by the Dutch lensgrinder and counterfeiter Zacharias Jansen in the early 1600s. Microscopes led to the discovery of the cell. (This term was first used by the British physicist Robert Hooke, but was in fact misapplied to the tiny *spaces* left by dead cells, which reminded him of prison cells.)

The discovery of sex cells (or germ cells) caused great excitement. Soon overenthusiastic microscopists were convinced that they had observed "homunculi" (tiny human forms) inside the cells, and it looked as if the problem of reproduction was solved. More importantly, the English botanist Nehemiah Grew speculated that plants and animals were "contrivances of the same wisdom." He suggested that plants too have sexual organs and exhibit sexual behavior. When the pioneer Swedish biologist Carl Linnaeus introduced his classification for species of plants and animals, the way was opened for more systematic research. The study of hybrids led to further speculation about the nature of genetic material.

For centuries it had been widely accepted that heredity was transmitted by "blood." (Hence the origin of such commonplace expressions as "blue blood," "blood line," "mixed blood" and so forth.) This was not only loose, but inadequate. How could the same parents produce differing offspring from the same "blood"? Also, what accounted for the appearance of characteristics not present in either parent, but seen in long-dead ancestors and distant relatives? For instance, in thoroughbred racehorse breeding, piebalds have been known to recur after a gap of *dozens* of generations. (This example reveals one of the great lost opportunities of genetics. All English thoroughbreds are descended from the forty-three "Royal Mares" imported by Charles

II, and three Oriental stallions imported a few years earlier. The breeding books trace each bloodline back to its origins, with notes on the characteristics of each progeny. Well over a century before genetics was born, any Newmarket trainer had at his fingertips sufficient material to found this science.)

By the mid-eighteenth century the scientists had at last started speculating along lines that were obvious to any racehorse breeder. The idea of evolution began to circulate. One of the early developers of this idea was the eighteenth-century philosopher-poet-scientist Erasmus Darwin (grandfather of the famous Charles). Erasmus Darwin was convinced that species were capable of change. Any creature with "lust, hunger and a desire for security" would organically adapt to its surroundings. But how?

The French naturalist Jean Lamarck came up with the first coherent theory of evolution. Lamarck had been born in 1744, the son of a broke aristo. By the age of thirty-seven he had become Botanist to the King. When the Revolution took place Louis XVI was executed, along with any blue-blood who could be found. But Lamarck quickly evolved a suitable social cover, and emerged as Professor of Zoology at Paris. In the light of such experience, it's not surprising that Lamarck believed in the effect of environment on evolution.

According to Lamarck, "acquired characteristics are inherited." In other words, a man who has learned how to become a skillful fencer will pass on this skill to his son. This sounds fairly plausible—especially when one considers the Bach family. A son often does exhibit certain characteristics acquired by his father. But not for Lamarck's reason. The fencer son may have inherited his father's athleticism and quick-wittedness, but not his *actual skill*. The fault in the "acquired characteristics" theory is demonstrated by a more extreme example: even after generations of being blinded at birth to work in coal mines, pit ponies were still not born blind. Nevertheless, not long after Lamarck died the idea of evolution gradually became more widespread. (To this day, there is a statue of Lamarck in the Luxembourg Gardens in Paris, inscribed THE INVENTOR OF EVOLUTION.)

The father of evolution received little recognition during his lifetime, but the father of genetics received none. Gregor Mendel was born in 1822 in Silesia, which was then part of the Austro-Hungarian Empire. His parents were peasants and he was forced to abandon his university studies because he had no money. In order to continue his education he entered the priesthood, where he taught himself science yet failed his simple teaching exams. Allegedly this was because of "examination amnesia," though the fact that he scored lowest marks in biology speaks of some more profound resistance to systemized knowledge.

Despite this, it was in systemization that Mendel showed his genius. Mendel ended up at a monastery just outside Bruno, in what is now the Czech Republic. Put to work in the monastery garden, he began a long and systematic series of experiments crossbreeding edible pea plants *(pisum)*. Mendel studied seven different characters of the plants, such as flower color, height, seed shapes and so forth. He found, for instance, that if tall plants were crossed with short plants, the result was tall plants. But when these first generation hybrids were crossed with each other, they produced 75 percent tall plants and 25 percent short plants.

Mendel concluded that each character was determined by two "factors," one contributed by each parent plant. For instance, the characteristic of height was determined by a "tallness" factor and a "shortness" factor. The "tallness" factor and the "shortness" factor both remained in the plants. They did not blend, they retained their separate identities—but one was dominant. In this case the "tallness" factor was dominant. This explained why when the plants were initially crossed, their hybrid offspring were all tall. But when the hybrids were crossed, the "tallness" and "shortness" factors split and reformed.

Each parent contributes one factor to each offspring, producing four possible combinations:

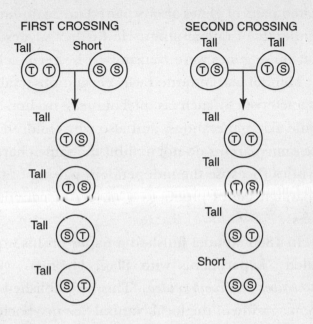

This accounted for the 75% : 25% distribution of tall plants and short plants after the second crossing.

Mendel's "factors" are basically what we now call genes. It looked as if genes held the key to heredity. After conducting over twenty thousand experiments, Mendel came to further conclusions. Firstly, plants inherited an equal amount of "factors" (or genes) from each parent. Also, sep-

21

arated pairs of genes always paired up again independently of one another. He further suggested that these genes were transmitted by germ cells.

Mendel had indicated why certain observable characteristics (such as piebaldness in horses) could skip generations, and also why children of the same parents do not exhibit the same characteristics (because the independent pairing of the separated genes produces a variety of combinations).

In 1866 Mendel finished a paper on his work called "Experiments with Plant Hybrids" *(Versuche über Pflanzenhybriden)*. This he published in the magazine of the local Natural Science Society at Bruno. The paper outlines Mendel's experiments and the brilliant statistical deductions which led him to his revolutionary conclusions. These conclusions—now known as Mendel's Laws—were to be the foundation of modern genetics.

But this lay years in the future. Not surprisingly, few leading scientists studied the pages of the Bruno Natural Science Society magazine. For the moment no one was interested in Mendel's revolutionary findings; so he sent his paper to the top German botanist, von Naegeli, at the University of Munich. Unfortunately Naegeli clung to the quaint belief in spontaneous generation. In his view, biological elements were created spontaneously by nature at cell level, and these elements then combined to produce pure species. The creation of species thus came about for no apparent reason, at the spontaneous whim of nature. According to this theory hybrids were mere freaks, and Mendel's experimental evidence was therefore irrelevant.

Despite Mendel's years of painstaking research, Naegeli told him that he needed to conduct yet more experiments if he was to convince anyone with his findings. Naegeli suggested that this time he should use hawkweed *(Hieracium)*. Sadly, hawkweed was an exceptional case, and Mendel's results did not match his previous findings. Mendel became somewhat disillusioned, and around the same time was elected abbot of his monastery. There was little time for further experiments on his previous exhaustive scale, and he died without recognition in 1884.

Not until 1900 did Mendel's work come to light. Only then, thirty-four years after the publication of his original paper, did he receive the universal acclaim he deserved. But such widespread renown can have its drawbacks. In 1936 Mendel's findings were scrutinized by the British scientist Sir Ronald Fisher, pioneer of modern statistics, who found that Mendel had committed the unforgivable scientific sin. On a number of occasions Mendel had fudged his figures to make his statistics fit his thesis. Fortunately the science of genetics was by this stage well and truly launched, with no chance of being sunk by this pedagogic depth-charge. (Modern genetics is not alone here. Margaret Mead, the mother of modern anthropology, established herself as the world leader in her field when she published *Coming of Age in Samoa* in 1928. Not until many years later, by which time anthropology had built a firm structure on this foundation, was it discovered that many of Mead's colorful and optimistic findings in this work were sheer fantasy. But anthropology, like genetics, was far too well-established to be sunk by mere facts.)

Mendel had conclusively disproved the "blood" theory of heredity—which implied that the characteristics of parents are blended in their offspring. But as his work remained unknown, this theory continued to flourish. Even Charles Darwin believed that heredity was transmitted in this fashion. He also accepted telegony—having witnessed a case where a mare, who had previously mated with a zebra, gave birth to a foal with stripes after mating with an Arab stallion. And unlike Mead or Mendel, Darwin had a scrupulous regard for the facts. We can only assume that he was hoodwinked by a cunning zebra owner, or one of the horses had a striped ancestor.

Fortunately Darwin's work in the related field of evolution was to prove more lasting. The publication of his *The Origin of Species* in 1859 introduced the idea of "the survival of the fittest." Species evolved by natural selection. The entire history of life on earth appeared to be explained.

Despite this, the French Lamarckists continued to believe in the inheritance of acquired characteristics. According to them, the giraffe had grown its long neck as a result of generations continually stretching for high leaves. This theory was to be conclusively disproved in the 1890s by the heartless German biologist August Weismann, who must have been deeply impressed by nursery rhymes in his childhood. Evoking scenes from "Three Blind Mice," he conducted experiments in which he amputated the tails of mice for several generations. Despite this grim practice, the mice's tails neither disappeared nor became shorter. Weismann drew an important conclusion: hereditary inheritance is carried by germ cells (sex cells), and is not influenced by what happens to the organism.

That other persistent myth, the blood theory, was finally laid to rest by Darwin's cousin Francis Galton. In another series of unfeeling but apparently vital experiments, Galton transfused blood from white rabbits into black rabbits. The rabbits may have felt as if they were turning green, but in fact the transfusions had no effect. When the black rabbits were well enough to resume their normal activities, it was found that none of their numerous progeny had white fur. Heredity was certainly not transmitted by blood.

Darwin may have explained what happened to hereditary characteristics, but how these were actually conveyed from generation to generation remained a mystery. Weismann and Galton had conclusively demonstrated that this happened at the cell level. More importantly, Mendel had shown that the information was carried by "factors" (genes)—but this information still languished in a back issue of the Bruno Natural Science Society magazine.

Meanwhile advances had been made in a field which for the moment appeared to have little relevance to genetics. In 1869 the twenty-five-year-old Swiss biochemist Friedrich Miescher was researching at Tübingen into the composition of white blood cells. For his source material he used bandages from the operating theater of a local hospital—a rich source of pus, whose main ingredient is white blood cells. By adding hydrochloric acid solution, he was able to obtain pure nuclei. He then stripped these down still further by adding alkali, then acid. In the process he obtained a gray precipitate quite different from any previously known organic substance. He named this "nuclein"—since it was part of the nucleus. This we now know as DNA.

Ten years later the German pioneer of cell-structure research, Walther Flemming, began using the newly discovered analine dyes to stain the nuclei of cells. He discovered that these dyes imparted color to a bandlike structure within the nucleus. This he named "chromatin" (from the Greek *chroma,* meaning color). A couple of years later it was discovered that nuclein and chromatin reacted in precisely the same way: they appeared to contain the same substance. Chromatin consists of what we now call (after it) chromosomes, which in turn contain nuclein—or DNA. And DNA is what makes up the genes discovered by Mendel. All the disparate pieces were beginning to come together.

However, we can only see this in hindsight. At the time, these developments were disparate. Those involved didn't know where their work was leading them—even if they did have immediate aims (such as discovering cell structure or understanding the patterns of heredity). Only when the connection between these developments was made would the further picture emerge.

As early as the 1870s, the German biologist Oskar Hertwig had made an important discovery while studying sea urchins under the recently developed light microscope. During fertilization sperm penetrated the egg, and the nuclei of the sperm fused with the nuclei of the egg. The importance of chromatin (chromosomes) in this fertilization process quickly became apparent when the Belgian embryologist Edouard van Beneden began studying an intestinal threadworm found in horses, called *Ascaris megalocephala*. This big-headed parasite had only a few, large chromosomes, which made for easier observation. Beneden found that the sperm and the egg both contributed the same number of chromosomes in the fertilization process. He also discovered that there is a constant number of chromosomes per cell, which varies according to the species. (The threadworm, for instance, has just four chromosomes per cell, whereas the human cell contains forty-six.)

But if the nuclei of the sperm and the nuclei of the egg both contained an equal amount of chromosomes, and both contributed an equal amount of chromosomes, the amount of chromosomes should *double* during fertilization. Beneden noticed that this did not happen. Instead the chromosome number remained constant, maintaining the characteristic number for the species. This process, by which the number of chromosomes halves in the germ cells (formed by the egg and the sperm), Beneden called "meiosis," from the Greek "to decrease." Meiosis was eventually explained by Flemming, the original discoverer of chromatin. He noticed that instead of merging directly, the chromosome groups split lengthwise into identical halves. These scattered through the cell, and *then* merged with each other. Here, at cell level, was a process which bore an uncanny resemblance to the splitting of "factors" described by Mendel.

During the early years of the twentieth century the American experimenter Thomas Hunt Morgan became aware of this resemblance; but he was unconvinced by Mendel's findings. Morgan, a great grandson of the man who had composed the American national anthem, undertook an exhaustive series of experiments breeding fruit flies *(Drosophila)*. These flies have a life cycle of just fourteen days, allowing for rapid statistical work. Despite encountering discrepancies with Mendel's findings (which were nothing to do with Mendel's occasional fudging), Morgan was eventually convinced that Mendel had been on the right track.

Extending Mendel's work on "factors" (genes), Morgan showed that *Drosophila* had four groups of linked genes. The fact that some genes frequently remained together from generation to generation suggested a linking mechanism. Morgan decided that they could only be joined together on chromosomes. As there were four groups of genes, he concluded that *Drosophila* had four chromosomes.

Further statistical work showed that the assortment of *Drosophila* characters did not follow Mendel's laws. This could be accounted for by the splitting and recombining of chromosomes which Flemming had already observed. The splitting allowed some genes on the same chromosome to reassort, whereas others remained linked. This meant that genes at a greater distance from one another on the chromosome were more likely to reassort. And the higher the frequency of reassortment, the further apart the genes. Morgan realized that genes could be mapped.

In 1911 Morgan produced the first chromosome map, indicating the relative location of five sex-linked genes. Just over a decade later he had extended this map to include the relative positions of over 2000 genes on *Drosophila*'s four chromosomes. Things were moving fast.

They began to move even faster when one of Morgan's students discovered a method of increasing the mutation rate of *Drosophila*. Hermann Müller discovered that when the flies were irradiated with X-rays, they produced mutations at 150 times their normal rate. They also produced mutations which didn't occur in nature. Weird hybrids with deformed wings and misshapen sexual organs began to appear. This led

Müller to conclude that X-rays caused a reaction between chemicals in the genes. Essentially, mutation seemed to be the result of a chemical reaction.

Müller's joy at this vital discovery was tempered by a grim realization. Science was moving forward without control. The legend of Frankenstein producing monsters in his laboratory was coming true. X-rays could also be used to produce mutant human beings.

Genetics was becoming aware of its inherent dangers. Discoveries in this field were discoveries about the secrets of life itself. They revealed how it passed from generation to generation, and how it *changed.* What was known could also be used.

For the time being the possibility of isolating the gene remained remote. All that scientists could observe, even through the most powerful microscope, was the dim shadow of the chromosome. Where genes were concerned, science was still feeling forward into the dark. But Müller's demonstration of how to increase mutation meant that the gene's properties could now be extensively analyzed. We might not have been able to see the gene, but we could find out what was there.

Müller's X-ray experiments made him famous, and in 1932 he took up a post in Berlin. A year later, a dangerous human mutation (not, as far as we know, arising from X-ray irradiation) took over the political reins in Germany. Neither Hitler's gene structure, nor his views on genetics, appealed to Müller, and he left the country. Alas, Müller merely exchanged the frying pan for the fire. He now moved to Stalinist Russia.

By coincidence, Müller here encountered the second extrascientific issue which twentieth century genetics would be forced to confront. Communism was creating the world of the future; social engineering was seen as a science—and vice versa. But things are not as easy as this.

Ultimately, the direction science takes will always be a matter of human choice. (We found out how to leave the planet, rather than clean up the mess we'd made of it.) Science may follow human wishes, but it does not *conform* to them. In communist Russia, it was expected to do so—at least where genetics was concerned.

Soon after Müller arrived, top Russian geneticists began to "disappear" because they did not subscribe to the prevailing theory. This was peddled by a crafty and ambitious charlatan called Trofim Lysenko, who claimed to believe in Lamarckism. The idea that the heredity of organisms (including human beings) could be influenced by their environment (such as society) had an obvious appeal to scientific thinkers of Stalin's caliber. Acquired characteristics (such as communist beliefs) could be inherited, and a new type of human being altogether would emerge in the coming utopia.

Lysenko's ideas were to render Russian biology a laughing stock for *thirty years* (1934–1964). During this period serious scientists were expected to believe that wheat raised under suitable conditions could produce rye seed, and similar tall tales. (By corollary, domestic pussycats evicted to live in the wild would produce tigers— which must have made Soviet citizens somewhat wary of stray cats.) Müller argued that such non-sense was utterly disproved by X-ray irradiation. Flies subjected to this also produced "natural" mutations—proving that they were the result of inner chemical changes, which had nothing to do with insect society. Müller eventually returned to the United States, where he became an active campaigner against the abuse of science, as well as its own abuses.

Heredity was transferred by chemical reaction, but how did this work? When analyzed, the gene-bearing chromosome was found to contain a number of different proteins and nucleic acids. Either one, or a combination of these, was evidently the carrier of genetic information. The proteins were the obvious choice, as they had a more diverse structure, and thus appeared capable of carrying more information.

This conjecture was disproved as a result of experiments carried out by two bacteriologists working on either side of the Atlantic. Back in the 1920s in London, Fred Griffiths had carried out experiments on pneumocci, the bacteria which causes pneumonia. Under the microscope, the surface of a colony of pneumocci cells appeared shiny and smooth when they were infectious, but when they were noninfectious the surface of the colony appeared rough. If the smooth infectious pneumocci were heated, they were killed, becoming rough and noninfectious.

When Griffiths injected mice with either noninfectious rough cells or noninfectious heat-killed smooth cells, the mice naturally remained unaffected. But if he injected the mice with living rough cells *and* heat-killed smooth cells, the mice *were* infected. When he examined these mice, he found that they contained infectious smooth cells. These had evidently reconstituted from a mixture of the two injected cells. Something in the dead cells had caused this transformation in the living ones. A nonliving constituent of the smooth cells was evidently capable of combining with an element of the rough cells. Further investigation showed that this change was permanent. It was inherited by the next generation of cells. Some nonliving chemical had transferred and altered the living gene.

The American bacteriologist Oswald Avery, working at the Rockefeller Institute in New York, set about trying to isolate this "transforming principle," as he called it. By 1944 he had shown that it was a nucleic acid. More specifically, it was deoxyribonucleic acid (known as DNA).

By this stage considerable progress had been made on the analysis of DNA, though without realizing its significance. In fact, just the opposite. This negative view of DNA was largely due to the Russian-born chemist P.A.T. Levene, who also worked at the Rockefeller Institute. Analysis had shown that DNA contained four bases: adenine, guanine, cytosine and thymine. These were arranged in varying order along a linking structure:

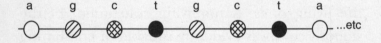

It was thought that genetic information would probably be carried by differing the amounts of each base. But Levene's state-of-the-art analysis indicated that DNA always carried equal amounts of the four bases. He concluded that DNA was a substance of boring structure and little significance. Proteins in the chromosomes were the carriers of genetic information, just as most suspected.

This view should have been exploded by the findings of his colleague Avery, which identified DNA as the "transforming principle." But Levene and Avery didn't get along. Temperamentally they were the tortoise and the hare. Levene had a striking, somewhat unsettling appearance: beneath his shock of hair his eyes were masked by tinted glasses. A headstrong workaholic, he was to publish an astonishing seven hundred papers during his scientific life—and saw himself as the genius-in-residence of the Institute. Avery, on the other hand, was temperamentally retiring: the son of a mystically inclined English clergyman. He worked with painstaking exactitude, and didn't believe in making a fuss over his findings. As a consequence, their importance was dismissed by the flamboyant Levene. To him, Avery's diffidence suggested that he remained unsure of his findings.

However, further analysis by Levene revealed that the nucleic acids had a much more complex structure than had originally been thought. DNA had a "backbone" consisting of sugar molecules (deoxyribose), linked by a bond (of phosphodiester). Attached to each sugar molecule was one of the four bases.

sugar ⟶

phosphate bond →

← base 1

← base 2

etc.

Such a molecule was very large, and was evidently capable of carrying genetic information. Reluctantly, Avery's findings had to be accepted. The tortoise had its part to play too.

At nearby Columbia University in New York, the Czech chemist Erwin Chargaff immediately embarked upon a further study of DNA. Using quantitative analysis, he discovered that different species each appeared to have their own characteristic DNA. Using the latest purification techniques, he managed to isolate the four nitrogenous bases: adenine, thymine, guanine and cytosine. By the early 1950s he had found that contrary to previous thinking, these four bases were not in fact precisely equal. Representing the bases as A, T, C and G, he found that:

$$A + G = C + T$$

also that:

$$A = T \text{ and } G = C$$

"Chargaff's rules," as these came to be known, would obviously be essential in future analysis of DNA.

But the fundamental question about DNA still remained. How did this "transforming principle" actually transform? In other words, how was the genetic information carried, and how was it conveyed? This was the "secret" contained in DNA: the secret of life itself, and how it passed on from one generation to the next. To understand this, it would be necessary to unlock the structure of DNA. This was the situation when Crick and Watson came on the scene.

Crick and Watson

EVEN AT SCHOOL, Francis Crick had his own attitude toward learning. He showed promise at mathematics, but was mainly interested in the answers rather than the means of reaching them. This attitude was to color Crick's entire approach to knowledge. With Crick you could always be sure of answers—lots of them, put forward with enthusiasm and conviction, even when they contradicted one another.

Francis Crick was born in Northampton in 1916, the son of a local shoe factory owner. He won a scholarship to Mill Hill, a small public school in the suburbs of London, and afterward studied at University College, London. Here he was taught the great scientific advances which had taken place at the turn of the century. Unfortunately he was aware that further advances had since taken place, rendering many of these redundant. Crick graduated with a second-class degree in physics and an attitude problem. Nowadays these twin attributes would disqualify him from further research—but Crick was not so easily discouraged.

Filled with belief in his own abilities, Crick applied to do research and was quickly allotted a task appropriate to the prevailing view of his character and abilities. His professor "put me onto the dullest problem imaginable": constructing a spherical copper vessel for testing water viscosity. Undaunted as ever, Crick remembers: "I actually enjoyed making the apparatus, boring though it was scientifically, because it was a relief to be doing something after years of merely learning." Crick had an independent mind, and was intent on doing something with it.

Crick was relieved from filling the world with

ball cocks by the outbreak of war. He was drafted to the Admiralty and put to work designing mines. In 1940 he married.

After the war Crick prepared to return to his research. In 1946 he attended a lecture by the American Linus Pauling, generally recognized as the finest chemist of the century. This awakened Crick to the possibilities of chemical research. Around the same time he also read *What Is Life?* by the Austrian physicist Erwin Schrödinger, one of the founders of quantum mechanics. This book suggested how physics, most notably quantum mechanics, could be applied to genetics. Although many of its brilliant suggestions were later "modified," even its errors were to prove inspirational to the coming generation of postwar scientists.

Organic molecules, the chemistry of genetics, quantum mechanics—this heady cocktail of research possibilities soon replaced the old ball cocks. 1947 saw Crick divorced and registered for research in Cambridge. Here he set about acquainting himself with the biological side of biological physics. Two years later he was taken on by the Cambridge Medical Research Council Unit at the world-famous Cavendish physics laboratory. Thus, at the somewhat mature age of thirty-three, Crick began his first real research work.

Undaunted by the fact that he only had two years of study in biology under his belt, Crick soon became renowned throughout the laboratory for his ability to produce a stream of innovative theories—usually concerning other people's research. Crick had at last found his vocation, and nothing could stop him. It quickly became obvious that an exceptional mind was developing—to say nothing of an exceptionally loud voice and a booming laugh. Some found his company refreshing, in limited quantities; others found his very presence gave them a headache. Among the latter was the head of Cavendish, the aging Sir Lawrence Bragg, who at twenty-five had been the youngest Nobel Prize–winner ever. A couple of years later a young American called James Watson arrived at Cavendish.

James Dewey Watson had been born in 1928

in Chicago. A child prodigy, he had been "discovered" by a local TV producer who put him on the "Chicago Quiz Kid Show." At the age of fifteen Watson was enrolled at the University of Chicago to study zoology. He wasn't keen on this subject (his real interest was in ornithology), and according to one of his teachers he remained "completely indifferent to anything that went on in class; he never took any notes and yet at the end of the course he finished at the top of the class."

At the age of nineteen Watson graduated and went on to the University of Indiana at Bloomingdale. Here he was affected by two crucial events. He too read Schrödinger's *What Is Life?*, which had a profound effect. The genius had discovered the gene, and he knew at once that this was his subject. But he was hardly qualified to pursue research in this area. As he admits: "at the University of Chicago I was principally interested in birds and managed to avoid taking any chemistry or physics courses which looked of even medium difficulty." With the blithe insouciance of youth (affecting both genius and dunce alike), "it was my hope that the gene might be solved without my learning any chemistry."

The second influential event in Watson's life at this point was studying with the microbiologist Salvador Luria, who had fled to America from Mussolini's Italy. Luria was a founder of the Phage Group, consisting of leading geneticists investigating self-replication at the viral level. Viruses were thought to be a kind of naked gene, and the simplest viruses are bacteriophages— often known simply as phages. Luria was making important advances in this field, using X-ray irradiation.

Schrödinger had shown Watson the direction, Luria showed him how to go about it. Watson launched into a doctoral thesis on phages, with Luria as his supervisor. Initially Luria was not bothered by Watson's lack of chemistry. Indeed, according to Watson: "he positively abhorred most chemists, especially the competitive variety out of the jungles of New York City." So Watson settled down to write a thesis on phages. However, by now Luria was beginning to wonder if the real nature of phages (and thus also of genes) would only become clear when their chemical structure was understood. So he suggested that Watson should try and pick up at least some chemistry.

Watson followed his mentor's advice with enthusiasm, and embarked on a do-it-yourself chemistry course. The results were spectacular, though not in quite the same way as his usual academic results. After Watson attempted to warm up some volatile benzine over a naked flame, he was no longer welcome in the labs. From then on, his chemical knowledge remained largely theoretical.

In 1950 Watson received a fellowship from the Merck Foundation, to study bacterial metabolism in Copenhagen under the supervision of the biochemist Herman Kalckar. A curious choice, considering his supervisor's profession. But Watson evidently thought otherwise: "Journeying abroad initially appeared the perfect solution to the complete lack of chemical facts in my head." Only when he found that Kalckar's English was completely unintelligible did he begin to have doubts about this enterprise. These deepened when Kalckar announced one day that his wife had left him and he was no longer interested in thinking about the digestive systems of bacteria.

Kalckar decided to get over things by spending a couple of months at the Zoological Station in Naples. He asked Watson if he'd like to come with him. This time Watson appears to have had no difficulty understanding his boss's English. He immediately wrote to the Merck Foundation for two hundred dollars in traveling expenses.

On an icy spring day in Copenhagen, the emotionally unbalanced biochemist and his nonchemist assistant set off for the sunny Mediterranean. This seaside break at the Merck Foundation's expense was to prove the most fortuitous inspiration of Watson's scientific life.

During Watson's stay in Naples, the city hosted an international scientific congress with "a small number of invited guests who did not understand Italian and a large number of Italians, almost none of whom understood rapidly spoken English, the only language common to the visitors." Here Watson met the thirty-three-year-old New Zealander Maurice Wilkins, who was based at King's College, London.

Wilkins had been a high-flying physicist, and during the war had worked in California on the Manhattan Project, which created the first atomic bomb. The result had left him disillusioned with physics, and after the war he had become interested in molecular biology. On returning to Britain, Wilkins had joined the Medical Research Council's biophysics unit at King's College. Here he had begun taking X-ray diffraction pictures of DNA. He had even brought one of these with him to Naples, and he showed it to Watson.

Wilkins's photo depicted a somewhat blurred geometric pattern, whose significance had to be pointed out to Watson. In a flash, Watson decided that this was what he had been looking for. This was the way to discover the chemical structure of DNA.

Despite knowing even less about X-ray diffraction than he did about chemistry, Watson wrote to the Merck Foundation demanding a transfer to the Cavendish Laboratory in Cambridge. Here the Medical Research Council had another X-ray diffraction unit, which had been recommended by Wilkins.

Copenhagen, Naples, Cambridge—all within the space of a year. The twenty-two-year-old whiz kid was certainly whizzing around. But what was he whizzing *about*? The Merck Foundation cut Watson's grant by a third, to two thousand dollars, and told him their support would end six months early, in May 1952 (just before the European summer touring season got under way).

The Foundation was determined that this time Watson should stay put. They needn't have worried. With the megalomaniac vision of youth, Watson had now seen precisely what he wanted to do. He would solve the secret of life, no less. He would discover the structure of DNA and become world-famous. This was ambition pure and simple. A few days after his twenty-third birthday the quiet, seemingly shy young Watson entered the Cavendish Laboratory in Cambridge.

It wasn't long before he met up with the owner of the famous laugh. The rapport with the thirty-five-year-old Crick was instantaneous. Watson was soon describing Crick as "no doubt the brightest person I have ever worked with and the nearest approach to Pauling [the great chemist] I have ever seen . . . He never stops talking or thinking." Crick appeared equally impressed by Watson: "He was the first person I had met who thought the same way about biology as I did . . . [he had] exactly the same ideas as I had, but I cannot remember in detail what they were." This is not surprising. At the time, Crick had only been studying biology for two years, while the young Watson already had a Ph.D. in the subject.

The gangling unsophisticated young American and the bumptious loudmouthed Englishman may have appeared different in many ways, but they undeniably had one thing in common: overweening self-confidence. The X-ray diffraction unit at Cavendish was studying the structure of protein; but Crick and Watson quickly decided that this was not the central issue. What *they* were interested in discovering was the structure of DNA.

Between the two of them, Crick and Watson mustered a considerable range of ignorance for this task. Crick had only two years of biology; Watson had little chemistry, and no X-ray diffraction experience. They were unlikely to be hampered by any excess intellectual baggage in their discussions.

These discussions soon began taking place on a regular basis. They would begin in the morning over coffee in their shared office. They would continue over lunch in the Eagle, a popular undergraduate pub, where Crick introduced Watson to the joys of warm flat English beer. And often they would even continue over dinner at Crick's tiny flat, where he lived with his new half-French wife, Odile. These conversations were not confined to Crick and Watson; they frequently involved any of their Cavendish colleagues who would listen.

Cavendish in Cambridge, along with King's College in London, were at the cutting edge of X-ray diffraction. Cavendish had already once changed the face of science. Several decades earlier Rutherford and his colleagues had founded nuclear physics, bringing this new science to fruition with a miraculous burst of creativity at Cavendish during the 1930s. Now it was the turn of molecular biology. And this was to be largely due to X-ray crystallography.

The head of Cavendish, Sir Lawrence Bragg, had played a leading role in the founding of X-ray crystallography, along with his father, Sir William Bragg. This was the technique which had enabled human vision to extend beyond the range of light. No matter how powerful a microscope is constructed, it can only see objects larger than the wavelength of light. X-rays are a form of electromagnetic radiation which has a wavelength 5,000 to 10,000 times shorter than the wavelength of light (which itself has a wavelength of $1/10,000$ or 10^{-4} centimeters). This makes the wavelength of X-rays similar in size to the distance between atoms in crystals.

When a fine beam of X-rays is passed through a crystal, the beam is diffracted by the atoms in the crystal and emerges as a complex pattern. If this pattern is recorded on a photographic plate, it is possible to deduce the structure of the crystal. This process may appear relatively simple, but it in fact involves a host of excruciatingly exacting and sophisticated techniques. These involve such tasks as positioning, refining and isolating the individual crystals, as well as attempting the deduction of highly complex molecular structures from dim patterns.

The X-ray crystallography unit at Cavendish was led by the Viennese-born biologist Max Perutz, who had left Austria in 1936. For several years Perutz's formidable experimental abilities, assisted by Bragg's equally formidable theoretical skills, had been devoted to determining the structure of hemoglobin (the protein of red blood cells). By 1951 they were at last beginning to achieve some success.

But Perutz and his team were not the only ones interested in this topic. The fifty-year-old master Linus Pauling was also trying to work out the structure of complex biomolecules. Working from his base at Cal-Tech (the California Institute of Technology), he had already deduced a model structure for proteins involving a helix—a spiral of molecules much like a corkscrew. He suggested that this might be the form of many

complex biological molecules, including DNA.
And in 1951, working from old prewar X-ray diffraction plates, he even went so far as to publish a
suggested structure of DNA, involving three
coiled helices.

At Cavendish, Crick and Watson studied Pauling's suggestion, but remained unconvinced.
Pauling simply hadn't filled in enough details.
His idea was really little more than a brilliant
hunch.

Meanwhile things were also progressing at
Wilkins's X-ray crystallography unit in King's College, London. Unlike our two free spirits at Cavendish, this was where the actual work on DNA
was supposed to be going on. (King's and Cavendish had a gentlemen's agreement: protein was
Perutz's baby, DNA was Wilkins's. But Crick and
Watson were far too interested in DNA by now to
worry about being gentlemen.)

Wilkins had by this stage been joined by the
twenty-nine-year-old Rosalind Franklin, who had
just completed four years of X-ray diffraction
work in Paris, and was very much state-of-the-art

in this new field. Franklin's arrival should have been a lucky stroke for Wilkins. She was both highly intelligent and attractive, even if she did choose to dispense with makeup and wear somewhat dowdy clothes. But this was 1950s Britain, very much a Stone Age where relations between the sexes were concerned. Quite simply, Wilkins had no idea how to deal with a woman in his laboratory. And "Rosy" Franklin was no ordinary woman. The willful daughter of a cultured Jewish banking family, she had her own ideas about how things should be run. Right from the start there was "chemistry" between the bachelor Wilkins and the unmarried Franklin. Unfortunately, it was negative chemistry. And to make matters worse, Franklin arrived under the impression that she was taking over the X-ray diffraction work on DNA. Wilkins, on the other hand, thought she was being taken on as his assistant. Wilkins and Franklin began working in difficult tandem.

As if all this wasn't enough, DNA was proving a particularly tricky subject for X-ray diffraction. It was a macromolecule, which had to be studied intact, as many of its most significant qualities were lost in any other form. Wilkins had received a particularly pure sample of DNA from Berne. This sample resembled treacle. As Wilkins explained, when a glass rod was raised from its surface, "an almost invisible fiber of DNA was drawn out like a filament of a spider's web." In this fiber individual molecules were aligned, and although DNA was not strictly crystalline, this didn't seem to matter. When much of the water was withdrawn from DNA, its structure exhibited orderly, repetitive, quasicrystalline qualities, which proved amenable to X-ray crystallography. This water-reduced form was known as "A-form" DNA, and was the sort initially used at King's.

By November 1951 Franklin had made significant progress. She had worked out a new method of reintroducing water to the A-form DNA. After rehydration the structure of the DNA was transformed. The differences showed up in X-ray diffraction patterns. Franklin had managed to ob-

tain some of the best pictures so far. Even so,
these remained very blurred, resembling a film
of a spinning four-blade propeller.

After measuring the angles and patterns that
could be deduced from the photographic plates,
Franklin began a mathematical analysis of the results. She soon came to some important conclusions about the overall structure of DNA.

Franklin decided to make public her findings

at a seminar in King's. Wilkins invited along Watson, knowing from their meeting in Naples that he was interested in DNA. (Though alas, Wilkins seems to have had no inkling of *how* interested Watson was in DNA—he didn't realize that Watson and Crick were planning to "scoop" him.)

So that he would be able to understand what Franklin was talking about, Watson hurriedly set about learning some crystallography (which is what he was supposed to be doing at Cavendish in the first place). He then set off for London to attend Franklin's seminar.

There he learned that Franklin's results seemed to confirm that DNA was helical. In her view, it consisted of anything from two to four interlaced helical chains. Each helix had a phosphate-sugar backbone, with attached bases (adenine, guanine, thymine, cytosine), much as Levene had suggested (see page 41). But importantly, it looked as if the bases were attached to the inside of the helix, possibly forming links between the helical chains.

After the seminar, Wilkins and Watson had a

Chinese meal together in Soho. Here Wilkins spilled the beans about the misery of life in the lab with Franklin. This was very much like an archetypical 1950s English marriage. Wilkins apparently withdrew into a shell of distant politeness, while Franklin adopted a cold insistent manner. There was precious little intercourse between them. To Watson, this didn't look like the team that was going to deliver the baby.

Watson returned to Cambridge on the night train in an inspired mood. Franklin didn't seem to be interested in trying to make a model of DNA. All she seemed intent on doing was making painstaking measurements of the diffraction plates, and trying to match them with known bond lengths between molecules. Her method was built on facts.

This just wasn't Watson's way of working at all. Following in the footsteps of his great compatriot Pauling, Watson believed in making models. Admittedly, this could be a bit of a hit-and-miss procedure. After you'd built up the jigsaw, the diffraction pictures often didn't fit. This meant twiddling a bit with the chemical bonds until they did. Here Watson's bible was Pauling's *The Nature of the Chemical Bond,* the greatest chemistry textbook ever written. This contained a blueprint for the structure of complex biological molecules at bond level.

Crick was another one who didn't believe in wasting time on unnecessary research. After all, theoretical speculation was his forte. (As they knew only too well in Cavendish: Crick was always poking his nose into other people's experiments, and coming up with instant theories. What was so galling was that his theories were usually brilliant—and sometimes even right.)

Unfortunately, Crick and Watson's model-building soon ran into a few local difficulties. For a start, it depended heavily upon Watson's suspect grasp of X-ray crystallography; in particular, his understanding of what Franklin had said at her seminar. Without further ado, Watson and Crick plumped for building a model with three interlinked helices. (After all, there was a three-to-one chance here.) But when it came to the question of whether to attach the bases to the inside or the outside of the helical chains, they definitely backed the wrong horse. They put the bases on the outside—presumably because Watson had forgotten, or misunderstood, what Franklin had said.

The trouble was, Franklin was dealing with the facts—and it was unwise to ignore these, if you wanted to come up with the right answer. Watson evidently didn't look at things quite this way. But Crick did. He was well aware of Franklin's devotion to the facts—but he had his own oblique view of this matter. He suspected that Franklin didn't know what she was doing. In his view, all the evidence necessary for determining the structure of DNA was quite possibly already in existence—lying among her diffraction photos.

Crick and Watson made an ideal Laurel and Hardy. Despite an age gap of twelve years, this was a partnership between equals. Both were brilliant at their chosen field (of which the other knew practically nothing). So neither felt be-

hoven to the other. Ignorant but original suggestions could be made, uncluttered even by the
misconceptions gained from half-knowledge.
And these could be dismissed by the other, without any hurt feelings. Yet there were moments
when such misguided suggestions could prompt
a hitherto unconsidered line of expert thought.
As a result, when Crick and Watson were good,
they were *very good*—and when they were bad,
they were laughable.

But they were aware of this (largely, one sus-
pects, because this was a permanent situation
with Crick). This was very fortunate, because the
model they originally came up with had precious
little to do with reality.

Oblivious to this state of affairs, Crick and
Watson proudly invited Wilkins and Franklin to
Cambridge for the day. They wanted the King's
team to look over their brilliant new model for
DNA. This was quickly exposed by Franklin as a
joke—though she herself remained unamused at
this waste of her time. Growing more irritated by
the minute, she fired off one question after an-

other, each appearing to expose a new flaw. The model just didn't fit the X-ray evidence. At all. Then it became clear that Watson had misunderstood something even more fundamental at Franklin's seminar in London. The A-form DNA which Franklin used was dehydrated. In order to build up the true structure of DNA, one had to allow for added water. Watson had done this, all right. But he'd got the figure wrong—woefully wrong. Their model had a *tenth* of the water it should have had.

The subsequent lunch in the Eagle was a sticky affair. Franklin was like a thundercloud; her unwilling partner Wilkins just wished he wasn't there; Crick attempted a little light bombast over his beer; and Watson sat squirming with uncharacteristic embarrassment over his glass of dry sherry.

By the time they returned to the lab, Crick felt more like his old self. In an ebullient mood, he refused to surrender without a fight. Watson gamely offered a little, rather lame, support, while the others stood in silence. Then Wilkins suggested that he and Franklin might be able to catch the early train back to London, if they hurried. The day was over.

Inevitably news of this debacle soon reached Bragg. The boss of Cavendish was also not amused. Bragg already had it in for Crick, whose very presence continued to give him a headache. Crick was quickly cast as the villain of the piece, leading the young American research student Watson astray. (If anything, it was of course the other way around.)

Bragg demanded to see Crick in his office, and launched into a firm dressing-down. Not only had Crick broken the gentlemen's agreement between Cavendish and King's, but he had endangered further government funding by the Medical Research Council, which funded both units. Times were still hard in postwar Britain, and many thought that the Medical Research Council was a waste of money. What was the point of the government funding purely theoretical science projects, such as research into the structure of protein and the gene, when the country had only just dispensed with food rationing?

Bragg began asking Crick a few pertinent questions. What the devil did he think would happen if word leaked out that King's and Cavendish were in fact duplicating their work, in an unnecessary competition to see who would "win"? Why, they'd all be out of a job. And if Bragg had anything to do with it, this situation would remain permanent in Crick's case. With the recommendation he was likely to receive from Bragg, he'd be lucky if he ended up researching the chemical properties of aspirins.

Bragg finished by expressly forbidding Crick from doing any further work on DNA. From now on, this would be King's' preserve. Crick was ordered to return to his work on protein, the work he was being paid for. Watson, for his part, was encouraged to return to his own field, phages. He chose to work on the structure of the tobacco mosaic virus (TMV).

That was the end of it. Just before Christmas 1951, the Crick-Watson race for the DNA title came to a full stop. Or so it appeared. But no one had reckoned with the sheer ambition of Crick and Watson, and the lengths they were willing to go to fulfill it.

In retrospect, it is reckoned that Watson had an "American" attitude toward ambition. On the other hand, Crick was in rebellion against the stuffiness of middle-class England (the discreet breeding ground of that now extinct species, the "gentleman"). Such attitudes would nowadays be largely welcomed, though at the time they were regarded as unscrupulous. And not without some justification, as we shall see.

Watson's project on TMV was in his own words "the perfect front to mask my continued interest in DNA." One of the main components of TMV was nucleic acid. In fact, its nucleic acid content was a variant on DNA called RNA, but Watson felt sure that this could "provide a vital clue to DNA." Crick's attitude was typically forthright. He may have been banned from *working* on DNA, but no one on earth could ban him from thinking about it.

Crick decided to try a new tack. Instead of the four different types of base being attached to the outside of the backbones of coiled helices (as in the previous model), they must in fact be on the inside. But how would there be room for this? Fortunately all four different types of base consisted of flat molecules. Crick decided (again based on no evidence whatsoever) that the bases must fit together like two interleaved decks of cards. In other words, they were stacked on top of one another inside the entwined backbones. And if they somehow attracted one another, this would also help hold together the exceedingly long thin threads of entwined helices (the backbones). Speculation piled on speculation like a house of cards.

As part of his new regime of thinking about (but not working on) DNA, Crick began speculating aloud with a few scientific pals over beers in the Eagle. Eventually he became deeply involved in a conversation with John Griffiths, a young mathematics postgraduate. John happened to be a nephew of Fred Griffiths, whose 1920s experiments on rough and smooth pneumocci had inspired Avery to prove DNA was the genetic carrier. This link was not entirely coinciden-tal. John Griffiths had a hunch that certain prob-lems of DNA could best be resolved by mathe-matics, and had already done a few preliminary calculations using known data about the four bases.

As ever, Crick was soon discussing the funda-mental problems. Any structure for DNA had to account for (or at least allow for) replication—the process by which it passed on its genetic in-formation. Crick knew that this must somehow involve the coded sequence of four bases, which now seemed to be stacked on the inside of the entwined helices.

Griffiths passed on to Crick some calculations he had done concerning the four bases—ade-nine (A), guanine (G), thymine (T) and cytosine (C). Griffiths had worked out which of the bases were attracted to one another. According to him, G attracted C, and A attracted T.

In a flash of supreme inspiration, Crick saw that this could be the key to DNA's replication. If the helical strands parted, they could then be-

come the templates for the formation of comple-
mentary strands *precisely similar to the ones from
which they had parted.*

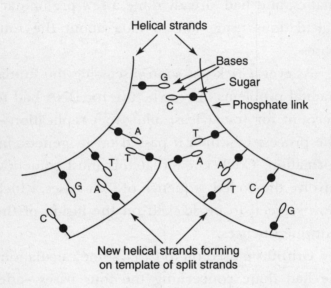

Helical strands

Bases

Sugar

Phosphate link

New helical strands forming
on template of split strands

This was indeed a giant leap of the imagination
on Crick's behalf—for he didn't understand that
Griffiths had envisaged a very different model of
his own. Griffiths had based his calculations on
the idea that the bases were lined up against each
other, edge to edge, and joined by hydrogen
bonds.

CRICK'S MODEL

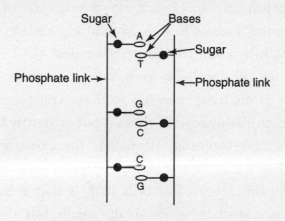

GRIFFITHS'S MODEL

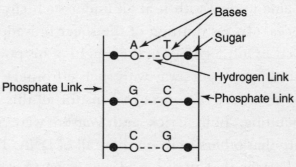

Another bonus of Griffiths's base attraction combination was that it at once accounted for Chargaff's rules (which decreed that the bases always occurred in the quantities A = T and G = C). Unfortunately Crick remained unaware of this clinching fact—for the astonishing reason that he hadn't heard of Chargaff's rules! (In Crick's defense, it has been pointed out that Watson must surely have referred to these rules at some stage, probably several times—but evidently Crick just wasn't listening. If such is the case for the defense . . .)

All this reveals not only Crick's staggering ignorance of what he was dealing with, but also his equally staggering powers of imagination in being able to deal with it at all under such circumstances. (To say nothing of the sheer bravado involved.) Only a genius at the height of his powers could hope to get away with such effrontery.

There were of course reasons for all this corner-cutting. Both Crick and Watson were well aware that others were on the trail of DNA. They felt certain that they would always be ahead of the opposition at King's (because Wilkins was still misguidedly keeping them informed of their progress). But Linus Pauling was another matter. He had already proposed one rather tentative structure for DNA. It was only a matter of time before he had a real go.

Then Watson learned that Pauling was due to arrive in London for a lecture. Inevitably, Pauling would want to see what was going on at King's. Previously he had only used prewar X-ray diffraction plates—but once he saw the latest plates there'd be no stopping him.

Crick and Watson could only grit their teeth, and go on pretending to work at their separate projects. The summer term had now started: Watson began playing tennis and taking an interest in girls. In order to differentiate himself from the other crew-cut Americans in Cambridge, he began anglicizing his Chicago accent and let his hair grow. Long hair was rare enough in the 1950s, but the results were rarer still in Watson's case. His bushy crop stood on end at great length, giving him an electrically shocking appearance. Only Crick didn't seem to notice, mainly because he was too busy laughing at his own remarks or drinking beer.

But the fates were on Crick and Watson's side. In May the news came through that Pauling wouldn't be coming to England after all. The world's greatest chemist had been prevented from boarding the plane at New York's Idlewild Airport. At the very last moment the U.S. State Department had withdrawn his passport on the grounds that he might defect to Stalinist Russia. Pauling was an outspoken advocate of a World Peace Conference, and this was tantamount to being a communist spy during the McCarthy era in America.

But it wasn't all good news. At King's, Franklin had made some spectacular advances in X-ray diffraction technique. These had now convinced her that DNA was *not* a helical structure after all. Even Wilkins seemed to concur with her view, albeit reluctantly. (As it later emerged, Franklin hadn't actually let Wilkins see her evidence at this stage. So one assumes it must have been the sheer force of her argument, or perhaps the force with which she put it, which carried the day.)

Watson now completed his work on TMV. According to him, the X-ray diffraction plates showed that this was a helical structure. In fact, his evidence was based on readings which Franklin (who was, after all, the expert) had now decided did *not* indicate a helix.

Despite Franklin's bombshell, Watson went on insisting that DNA *must* be helical. He was encouraged by Crick, who did know what he was talking about. By now Franklin had allowed Wilkins to study her pictures, and he had shown these to Crick and Watson when they visited him in London. One glance, and Crick had decided that Franklin's non-helical theory rested on a misinterpretation. Although the pictures did not show the radial symmetry necessary for helices, in his view this was due to overlapping patterns of crystals. Crick's was a brilliant and daring conjecture—which had the added advantage of agreeing with what he thought was the case. So Crick and Watson remained unconvinced by the world's leading DNA X-ray crystallographer.

Other opinions were not so easily dispensed with, however. In July 1952 Chargaff himself arrived in Cambridge. Crick and Watson badgered John Kendrew, Perutz's brilliant assistant, into arranging a meeting with the great man.

Initially Chargaff was wary. Who were these interlopers, who claimed to know so much about DNA? He was one of the world's leading experts on the subject, and he had never even heard of them. When Chargaff was informed that the young man with the Harpo Marx hair and fake English accent was in fact an American, he understandably decided that he was in the presence of a nut. (In his disarmingly frank account of this period, Watson is not afraid to appear the fool. Yet it is worth bearing in mind, this twenty-four-year-old who was playing the fool was also playing an equal role in one of the major scientific breakthroughs of all time.)

To begin with, Chargaff wasn't quite so certain about Crick. But Crick soon provided him with sufficient evidence to judge his caliber. One can only imagine Chargaff's reaction when Crick unwittingly let slip that he hadn't heard of Chargaff's rules. In order not to waste any further time, Chargaff began asking Crick a few basic questions. In Watson's words, "he led Francis into admitting that he did not remember the chemical differences among the four bases." Unabashed as ever, Crick explained "he could always look them up."

Several years later, Chargaff was to write bitterly: "Mythological or historical couples—Castor and Pollux . . . Romeo and Juliet—must have appeared quite differently before the deed than after. In any event I seem to have missed the shiver of recognition of a historical moment: a change in the rhythm of the heartbeat of biology. So far as I could make out they wanted, unencumbered by any knowledge of the chemistry involved, to fit DNA into a helix."

Crick and Watson had done it again. But there was evidently something endearing about this pair of comedians. Even if their enthusiasm may have appeared misguided to many, it was certainly infectious to some.

In the autumn of 1952 Watson made friends with Linus Pauling's son Peter, who had arrived at Cavendish to do postgraduate research. Peter Pauling was invited to share Watson and Crick's office, and was soon enthusiastically joining in their conversations.

One day Peter Pauling informed Crick and

Watson that he had received a letter from his father. Grimly they listened as Peter Pauling told them that his father had once more turned his attention to DNA. He was putting together a paper outlining its structure, and had promised to send Peter an advance copy.

So that was it. Even if they'd wanted to, Crick and Watson knew that they couldn't compete with Linus Pauling. Not under the present circumstances. Any serious attempt on their behalf to arrive at the enormously complex overall structure of DNA relied upon model building, and that was now out of the question. (The terms of Bragg's ban were particularly explicit on this point.) All they could do was speculate on what Pauling might come up with—and this soon proved too galling for words. From then on, the conversations in the office were largely limited to Watson and Peter Pauling—who shared a common enthusiasm for the local Danish au pair girls.

Peter Pauling duly received a copy of his father's paper. After reading it, he passed it on to Crick and Watson. They read that Pauling senior had come up with a structure containing three helically entwined chains with the sugar phosphate backbone *outside* the coil. This was uncannily similar to the one which Crick and Watson had shown to Franklin and Wilkins on their disastrous day trip to Cambridge—except that Pauling had paid a little more attention to working out the details, and matching these to X-ray evidence (to say nothing of making sure that he included the right amount of water).

Despairingly, Watson "wondered whether we might already have had the credit and glory of a great discovery if Bragg had not held us back." Even in despair, Watson was still capable of exceptional feats of optimism. But this was only the start. Watson now persuaded himself that perhaps the world's greatest chemist had made a mistake. Suppose he'd botched one of his sums, or made an error in the chemical bonds?

With the obstinacy of youth, Watson settled down to check the precise details of Linus Pauling's structure—the chemical bonds, the figures, the location of key atoms. "At once I felt that something was not right." And for once *he* was right. Unbelievably, Pauling had omitted to give the phosphate groups, which formed the links in each chain, any ionization. This meant there was no electric charge to hold together the long thin chains. Without these they would simply unravel and fall apart. Worse still, without this ionization, the model which Pauling had proposed for this nucleic acid *wasn't even an acid*. The world's greatest chemist had made a basic schoolboy howler. (This was even better than Crick and Watson's efforts in this field—the absence of water molecules, the ignorance of Chargaff's rules, etc.)

It was obviously only a matter of time before Pauling realized his blunder. Crick and Watson reckoned they had just six weeks to come up with an answer of their own.

The twenty-four-year-old Watson couldn't contain himself over spotting Pauling's gaffe. After telling anyone in Cambridge who would listen, he set off on the train to London, so he could tell them at King's. Wilkins was busy when he arrived, so he burst in on Franklin in her lab (normally a holy of holies where few dared to venture). He at once showed her Pauling's paper, pointing out the error he had noticed. Unfortunately he also felt the need to point out how the great chemist's three-helix DNA structure bore an amazing resemblance to the one which fifteen months previously he and Crick had proposed (and she had so taken against).

This was an unwise move. Franklin did not take kindly to snide criticism, especially from the likes of Watson (whom she understandably viewed as a conceited young man whose pretensions were rivaled only by his incompetence). Icily containing her anger, Franklin reminded Watson that there wasn't the slightest evidence to support a helical structure for DNA. Watson foolishly began contradicting her, citing evidence from his own work on the tobacco mosaic virus (TMV). This eventually provoked Franklin to such an extent that she emerged from behind her lab bench—apparently bent on attacking him. As she advanced across the lab, the door opened. Wilkins had arrived, just in the nick of time. Watson fled through the door and escaped down the passage. (Discussions about the key to life are evidently just as dangerous in a lab as they are in a pub.)

Afterward, Wilkins comforted the shaken Watson as best he could. He even went so far as to show Watson some of Franklin's latest X-ray work. This was truly amazing. Franklin had managed to obtain X-ray diffraction pictures of an entirely new form of DNA. This B-form, as it became known, occurred when the DNA molecules were surrounded by large amounts of water. This produced X-ray diffraction patterns of astonishing clarity and simplicity.

"The instant I saw the picture my mouth fell open and my pulse began to race," Watson remembered. It was unbelievable that Franklin was still sticking to her non-helical theory. Admittedly, the A-form DNA evidence was ambiguous; but this new B-form left no doubt whatsoever (in Watson's view). These pictures showed that DNA was unmistakably helical in form. And their amazing clarity pointed to even more exciting conclusions. After a few minutes' calculations, it should even be possible to work out *how many* helical chains there were.

Sitting in the freezing railway carriage on the return journey to Cambridge, Watson excitedly began making sketches and calculations on the edge of his newspaper. By the time he went to bed that night he had decided that DNA consisted of two interwoven helical strands.

Next morning Watson was hyper. The moment he burst into Cavendish no one was safe from his latest ideas about DNA. When Bragg unwisely stepped outside his office, even he received an ear-bashing on the subject. Instead of going ballistic at this direct contravention of his order, Bragg was surprisingly sympathetic. Quite unexpectedly, he even gave Watson permission to construct a new DNA model in Cavendish. The way he saw it, there was no longer any gentlemen's agreement with King's—the main competition was now Pauling. (Also, Watson had craftily given the impression that he was working on his own. Bragg was not yet aware that he had just sanctioned another bout of Crick at his decibel-rich best.)

The machine shop downstairs at Cavendish

was put into immediate production of metal plates, in shape and scale-size corresponding to the four bases. In no time Crick and Watson set about building up a scale model, fitting together the intricate structure of two interlinking helical chains of molecules. If their hunch about Pauling was correct, they now had just three weeks left to come up with an answer.

Everything had to be built up from the basic building blocks of the known chemical contents of the complex DNA molecule. The size of each of the individual molecules which combined to form this complex molecule, and the lengths and angles of chemical bonds between them, all had to be taken into account.

An idea of the sheer mind-boggling complexity of this task is given by the following analogy. Imagine a couple of combs, both two meters in length, with uneven teeth sticking out at odd angles. These combs must both be twisted into corkscrews, and then intertwined, so that each tooth on one comb meets up with the complimentary tooth of the other comb. But before even beginning, it is necessary to calculate the exact length, position and angle of each individual tooth of each comb.

An idea of the scale involved is given by the fact that the combined coiled width of the two combs is less than two nanometers. (A nanometer is 10^{-9} meters, in other words one billionth part of a meter.)

As we have seen, Crick and Watson had already put considerable thought into these matters. But other features had to be accommodated too. An important factor was the precise twist of each helical chain of molecules—whether it was coiled closely like a spring, or more openly like a spiral staircase. Watson had surmised from Franklin's X-ray diffraction pictures of B-form DNA that the structure was a double helix, but her data also provided a few even more essential clues. For instance, it was possible to gauge from the patterns on the X-ray plates the exact diameter of the molecule (in the region of 1.6 nanometers). The angle of the ascending "screws" of the helices and how far they rose in a complete "circuit" could also be calculated with a much greater degree of certainty.

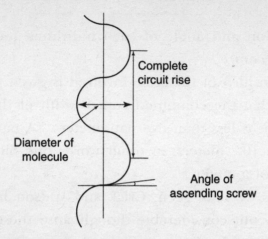

Complete circuit rise

Diameter of molecule

Angle of ascending screw

Gathered from X-ray diffraction picture
such as below

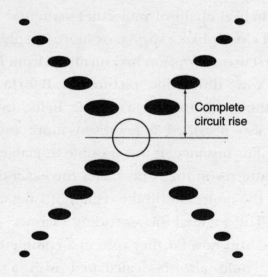

Complete circuit rise

Franklin's new clearer data also meant that Crick and Watson found themselves in an unusual situation. These were precise *facts,* which *had* to be taken into account. If their model didn't fit them, it was they who were wrong, not Franklin. And no amount of fudging could make it otherwise.

Not surprisingly, a few errors were made to begin with. And not surprisingly, given the participants, these were occasionally basic ones.

Against their own previous judgment, Crick and Watson were at first inclined to follow Pauling's suggestion that the bases existed on the *outside* of the entwined helical chains. Luckily, they were both possessed of sufficient self-confidence to soon abandon this conjecture. They had been right; and the greatest chemist in history was once again wrong. The bases had to be on the inside.

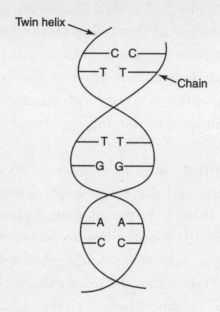

Griffiths had suggested that these bases were attracted to one another C to T, and G to C. But wouldn't it be better, and much simpler, if the bases were attracted like to like? This made for much easier formation of new molecules when the chains split to replicate. It seemed the ideal solution.

Then Watson discovered that the different like-to-like combinations (C + C, G + G, etc.)

were all different sizes. These combinations simply wouldn't fit inside the coiled chains of two regular helices. After a few further calculations he made an even worse discovery. This appeared to be true of *any* combination of bases. None of them fit inside the coiled chains. Pauling seemed to be right: the "bases inside" idea just didn't work. By this stage their painstakingly constructed model was already half made, but they were going to have to abandon it and start all over again. The trouble was, they now just didn't have time to build another model.

Crick refused to give up. It just didn't make sense to put the bases on the outside. Hour after hour he continued obsessively fiddling with the model, measuring the length of the bonds again and again, trying to reorganize them so that they fit inside the chains. As we know, Watson was not one to despair either—but his reaction to the crisis was somewhat different. The tennis season had begun; and also there was a new influx of Scandinavian au pair girls turning up at the parties.

Crick became increasingly irritated as Watson flitted in and out of the lab, pointing out how Crick's "inspired" suggestions wouldn't work, before disappearing in search of more congenial bonding arrangements. But Watson was also following his own lines of thought, albeit on a more sporadic basis. In the course of these he made an important discovery. He and Crick could have been making their calculations for the wrong isomeric form of the bases. This was not such a basic error as it might seem. Each base had a molecular formula which allowed for two different molecular structures—the enol form, and the keto form. All the evidence had pointed to the enol form—was it possible this was wrong?

Watson plunged into some lightning calculations, but to no avail. Even in the keto form, when the base pairs joined like-to-like they still didn't fit the chain. Then he discovered that when the keto-form base pairs joined A-T and C-G, just as Griffiths had suggested, they *did* fit inside the chain. And what was more, when they joined in this fashion the two different pairs were identical in shape and size. This meant that either pair could occur anywhere in the chain, thus allowing for a vast permutation of pairs. They'd done it! At last they'd discovered the key to the structure of DNA.

After a series of frantic readjustments, and the odd bit of fine tuning, the model was complete. On March 7, 1953, just five weeks after they had started building, Crick and Watson proudly unveiled their model to their colleagues at Cavendish. Word quickly began to spread around Cambridge. Within a few days the rumor had filtered out into the academic world at large. Some boffins in Cambridge had discovered the secret of life.

On April 25, 1953, Crick and Watson published a paper in *Nature,* unsensationally titled "Molecular Structure of Nucleic Acids." It said all it needed to say in just nine hundred words and a simple diagram.

Wilkins was characteristically unselfish in defeat, writing jauntily to Crick and Watson: "I think you are a couple of old rogues . . ." Others were to be less charitable about Crick and Watson's unscrupulous use of material from

King's' X-ray diffraction unit. In their view, Crick and Watson had no right to claim for themselves the credit for this momentous discovery.

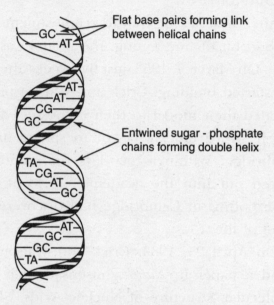

A section of the final DNA molecule arrived at by Crick and Watson: "the structure resembles a spiral staircase with the base pairs forming the steps."

Such views were taken into account by the Nobel Committee. In 1962 the Nobel Prize for Medicine was awarded jointly to Crick, Watson and Wilkins. Sadly, Rosalind Franklin had died of cancer four years previously at the age of thirty-seven. To emphasize the joint nature of DNA's discovery, and the assistance given to Crick and Watson by colleagues at Cavendish, the Nobel Prize for Chemistry in the same year was awarded

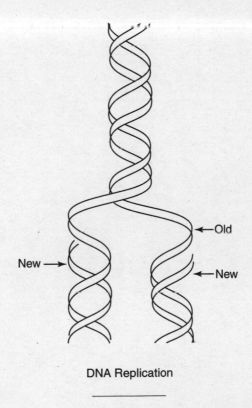

DNA Replication

to the head of the Cavendish X-ray diffraction unit, Max Perutz, and his colleague John Kendrew. Even so, it is Crick and Watson who will forever be linked with the discovery of the structure of DNA.

Afterword

THE CLOSE relationship between Crick and Watson began to founder somewhat once the public accusations started to fly. Watson soon returned to America, while Crick continued to work at Cambridge. Crick was to remain there on and off for over twenty years, becoming a leading force in the newly opened field of molecular biology. His main work was on DNA replication and how genes carry particular information. He has done much pioneer work on how to crack the "code" of the DNA bases.

In 1977, at the age of sixty-one, Crick moved

to California where he worked at the Salk Institute in San Diego. Crick has also managed to keep up a steady flow of typically "brilliant" ideas. In 1981 he published a book called *Life Itself,* in which he argued that life on earth originated from outer space. His theory is every bit as off the wall as it seems. (Unmanned rocket arrives from distant part of the galaxy, carrying primitive spores from a super-civilization which evolved billions of years ago. Now read on . . .)

Meanwhile, back on planet earth, Crick has also come up with perceptive ideas on the increasingly central question of consciousness (what is it? how does it function? do animals and plants possess it? etc.). Crick is still very much alive and laughing.

Watson's post-discovery career has been a similar rollercoaster. Back in America he soon took up a prestigious post at Harvard, where he continued to do research into DNA (particularly its role in synthesizing proteins). In 1965 he published *Molecular Biology of the Gene,* which is widely regarded as the best textbook of its kind.

Three years later he published *The Double Helix,* his personal account of the discovery of DNA. Many saw this as an attempt to recapture the limelight. In this he certainly succeeded. His treatment of Rosalind Franklin in this work ensured a public controversy. Yet the book also proved a classic: the finest autobiography of scientific discovery ever written.

In 1988 Watson went to Cold Spring Harbor on Long Island. Here he ran the Human Genome Project, whose aim is to map all the 100,000 human genes (DNA double helices, estimated to contain in all some three billion base pairs). Despite proving a brilliant administrator, Watson departed from this project with some acrimony in 1993. Sources differ as to whether he jumped or was pushed. The official version is that he resigned on a matter of principle—because he opposed the idea of patenting genetic information from the project. But according to no less a source than the *Encyclopaedia Britannica,* he "re-

signed because of alleged conflicts of interests involving his investments in private biotechnology
companies.'' Either way, when you make one of
the major discoveries in the history of science at
twenty-five, anything afterward is bound to seem
an anticlimax.

Genetics: A Few Facts, Fantasies and Fizzles

THE DISCOVERY of the structure of DNA founded an entirely new branch of science—molecular biology. This has resulted in a firework display of new human knowledge. The fuse lit by molecular biology soon exploded into a burst of original technologies and fields of research. Many of these—such as gene cloning, gene banks and DNA identification—were literally inconceivable just a few decades ago. And it's a racing certainty that by the turn of the century the Human Genome Project, or some other branch of DNA research, will produce another of the twentieth century's most important discoveries. What this will be, we don't know. Yet it's more than likely that it will revolutionize our entire understanding of what it means to be a human being.

• The Italian nuclear physicist Enrico Fermi once speculated along the following lines:

Our galaxy contains 10^{11} (i.e., 100 *billion*) stars, and there are at least 10^{10} galaxies. In the 10^{10} years since the universe began, many of these must eons ago have developed highly intelligent life forms capable of space travel. The earth is particularly favorable to such creatures.

"They should have arrived here by now, *so where are they?*" Fermi asked his Hungarian colleague Leo Szilard.

Szilard replied: "They are among us, but they call themselves Hungarians."

* * *

• Already we can see *how* to play God: by "constructing" the DNA structure for any type of individual. One day, perhaps sooner than we think, we might be able to *do* this. But even looking on the bright side, this leads into a paradoxical unknown. We could certainly eliminate disease. We could also aim to produce exceptional individuals—a Picasso, say, or the next Einstein, or even another Crick. A genius is by definition the most individual of individuals (deriving from the Latin *genius:* that particular quality which is innate to a person or thing). If we can make one, we can clone it. But then individuality ceases to exist . . .

And this is looking on the bright side.

• When asked what the study of biology had told him about God, the geneticist J.B.S. Haldane replied: "I'm really not sure, except that he must have been inordinately fond of beetles." There are over 300,000 species of beetles, all of which are truly differentiated species consisting of individual highly complex organisms. By contrast, there are only 10,000 species of birds.

* * *

• Concerning the Human Genome Project, Watson wrote in 1990: "The United States has now set as a national objective the mapping and sequencing of the human genome." A genome is all the genes contained in a single set of chromosomes, such as a parent donates to its offspring. There are twenty-three human chromosomes in a human sex cell. Each chromosome contains around 100,000 genes, or DNA double-helices. This combined collection of DNA helices contains around three billion base pairs.

Watson likened this project to the attempt to put a man on the moon. It is equally ambitious, and will cost a lot less. It is also likely to prove of infinitely more worth to humanity—unless this species is intending to leave despoiled planet Earth for purple pastures elsewhere.

• There are over four thousand inherited human disorders, resulting from genetic defects. These range from sickle cell anemia to Huntington's disease; and genetics may even play a part in Alzheimer's disease and certain types of schizophrenia. Such diseases can only be treated. At present no inherited disease can be cured.

As parts of the human genome become mapped, we are learning how to change the structure of the gene. This will enable us to prevent such diseases, and much more.

• If homosexuality is the result of an inherited gene pattern, and this can be altered to result in heterosexuality, should this be done? (Imagine vice versa.)

• Entries in scientific encyclopedias now include: gene amplification, gene bank, gene cloning, gene expression, gene imprinting, gene mutation, gene pool, gene probe, gene sequencing, gene splicing, gene therapy, gene tracking, genetic code, genetic engineering, genetic fingerprinting, genetic mapping . . . These are all happening *now*.

• When the Human Genome Project was set up in the 1980s, it was estimated that it would take until the middle of the twenty-first century to complete. With the advances in computer technology, this was later revised to 2015. Many now believe that this project will be completed before the turn of the century. (The French claim they have done it already, but refuse to let anyone see the evidence in case they pinch it.)

• We know that genetically engineered freaks have already been produced. Genetically enhanced vegetables are already on sale in supermarkets. Farm animals have been "improved" with a view to better meat production. More disturbingly, a mouse with a human ear growing out of its back has also been bred. And these are just the things we *know* about . . . Frankenstein-type experiments are no longer just fiction. As the molecular biologist John Mandeville recently put it: "We will be able to produce almost anything except a genetically engineered winning lottery ticket."

• We only know the function of the 2 percent of the human genome which contains genes. The purpose of the other 98 percent remains unknown—the biggest unsolved mystery of molecular biology. One suggestion is that this is a junkyard for discarded genes, which could even prove amenable to a form of genetic archaeology. (This would allow us to see what we, and indeed life itself, *might* have become.) Another suggestion is that this "empty zone" is in fact a breeding ground for entirely new forms of genes, and thus provides a kind of ghostly hand on the tiller guiding the direction of life.

Dates in the History of Science

pre 500 B.C.   Pythagoras discovers his theorem

322 B.C.   Death of Aristotle

212 B.C.   Archimedes slain at Syracuse

47 B.C.   Burning of Library at Alexandria results in vast loss of classical knowledge

199 A.D.   Death of Galen, founder of experimental physiology

529 A.D.   Closing down of Plato's Academy marks start of Dark Ages

1492   Columbus discovers America

1540	Copernicus publishes *The Revolution of the Celestial Spheres*
1628	Harvey discovers circulation of the blood
1633	Galileo forced by Church to recant heliocentric theory of solar system
1687	Newton proposes law of gravitation
1821	Faraday discovers principle of the electric motor
1855	Death of Gauss, "prince of mathematicians"
1859	Darwin publishes *Origin of Species*
1871	Mendeleyev publishes Periodic Table
1884	International agreement establishes Greenwich meridian
1899	Freud publishes *Interpretation of Dreams*
1901	Marconi receives first radio transmission across Atlantic

1903	Curies awarded Nobel Prize for discovery of radioactivity
1905	Einstein publishes Special Theory of Relativity
1922	Bohr awarded Nobel Prize for Quantum Theory
1927	Heisenberg publishes Uncertainty Principle
1931	Gödel destroys mathematics
1937	Turing outlines limits of computer
1945	Atomic bomb dropped on Hiroshima
1953	Crick and Watson discover structure of DNA
1969	Apollo 11 lands on the moon
1971	Hawking proposes hypothesis of mini black holes
1996	Evidence of life on Mars?

Suggestions for Further Reading

Francis Crick: *What Mad Pursuit* (Weidenfeld & Nicholson)—A personal view of scientific discovery.

Robert Olby: *The Path to the Double Helix* (Dover)—A broader, alternative view.

Anne Sayre: *Rosalind Franklin and DNA* (W. W. Norton)—Written by one of Franklin's close friends, this offers an alternative view of Crick and Watson's competitor.

James Watson: *The Double Helix* (W. W. Norton)— Best firsthand autobiography of scientific discovery ever written, filled with personal details as well as science. Biased (against Franklin, of course), but a great read for scientist and non-scientist alike.

Watson and others: *Recombinant DNA* (W. H. Freeman)—A "short" course (626 pages) in the basics.

NEWTON AND GRAVITY

A good case can be made for Isaac Newton being the finest mind humanity has yet produced. His theory of gravity offered his contemporaries their first glimpses of how the universe actually works, and his mathematics enabled later generations to walk on the moon. Today, we know that gravity is what keeps our feet on the ground, but how many of us know how Newton's greatest discovery really works? In *Newton and Gravity* Paul Strathern encapsulates several of Newton's more world-altering discoveries, explaining in lively prose their cultural context as well as Newton's early obsession with science (bordering on dementia), which made his revolutionary vision possible. Just a few of the big ideas covered here are:

- Newton's discovery of calculus at age twenty-three
- Why one of the greatest human insights of all time was in fact a hunch, and how it actually works
- Why it took Newton twenty years after his discovery

to reveal to the world the secret of gravity and planetary motion

Ideal for the intelligent reader eager to understand better how and why the universe works the way it does, *Newton and Gravity* is a fascinating refresher course that makes physics not only fun but shockingly easy to understand.

HAWKING AND BLACK HOLES

Stephen Hawking is arguably the most famous scientist in the world, and many of us know that black holes are his forte, but do we really have any idea what a black hole is? In this remarkably engaging book, Paul Strathern not only demystifies Hawking's universe-expanding theories, but helps readers appreciate why such knowledge is essential for all those who want to understand more fully the world around them. Just a few of the big ideas featured in *Hawking and Black Holes* are:

- How the universe originated, and what this has to do with black holes
- How the Big Bang actually worked
- Why black holes aren't actually black
- Is a Unified Theory of Everything possible (the ultimate big idea)?

Hawking and Black Holes also portrays the iron-willed determination of a man who continues to search for the key to understanding the cosmos, despite the devastating effects of motor neuron disease. Brilliantly simplifying the most complex ideas, *Hawking and Black Holes* will help you grasp the universe in ways you never thought possible.

EINSTEIN AND RELATIVITY

$E = mc^2$: Few equations have entered our consciousness with the speed and impact of Einstein's cosmos-changing formula. From the moment he published his revolutionary papers on his theory of relativity, humankind's view of the world and the universe changed forever, the latest phase of the modern age was born, and our horizons shifted.

We may know Einstein was the epitome of genius, but how many of us know what his theory really means, and what its realistic implications are? *Einstein and Relativity* presents a brilliant distillation of Einstein's life and work, and his work's historical and scientific context; and offers a truly accessible explanation of the concept that shaped the twentieth century.

Just a few of the big ideas covered here are:

- Einstein's discovery that light is both a particle and a wave

- How Einstein proved the existence of molecules
- Why there is no such thing as real time
- How Einstein's brilliance led to his worst nightmare—the atom bomb

Revealing Einstein's legendary eccentricities and his entry onto the world stage, this witty, engaging, accessible book is for anyone fascinated with this quirky icon, and even more important, for all those who want to truly understand the discoveries that have defined the modern world we live in.

TURING AND THE COMPUTER

The computer has revolutionized the modern age of communication, touching every part of modern life. Without a doubt, the development of the computer was a massive leap forward in humankind's progress and will stand as one of the twentieth century's greatest achievements. But how many of us know how it really works?

Turing and the Computer offers a brilliant encapsulation of the groundwork that led to the invention of the computer as we know it, and an absorbing account of the man who helped develop it. Eccentric and principled, Turing would lay aside a brilliant career in mathematics to serve his country by breaking German codes during the Second World War. Openly homosexual, he would later be put on trial on indecency charges and forced to undergo hormone treatments that wrecked his body and spirit. But the modern machine he helped create lives on.

Just a few of the big ideas included in this riveting book are:

- How Turing mapped out the theory of computers before a single computer had been conceived
- How Turing's Colossus broke the German Enigma codes
- Turing's proof of the existence of artificial intelligence

Concise and thoroughly compelling, *Turing and the Computer* is for all those curious about the philosophy and mechanics behind the now indispensable computer, and for anyone awed by the spark of invention that inspired its birth.

CURIE AND RADIOACTIVITY

Long upheld as a standard of independence and achievement in a time of extreme chauvinism within the scientific community, Marie Curie has often been cast as a secular saint who sacrificed her life for her contributions to science. But Marie Curie was no pedestal-riding Victorian effigy. Her scientific collaborations with her husband, Pierre, and later her Nobel Prize–winning daughter, Irène Joliot-Curie, are among the most innovative and productive partnerships of the century, resulting in major advances in nuclear physics and treatment for cancer and other diseases. And when Marie Curie's passionate love for science sparked an affair with the married French physicist Paul Langevin, her international fame turned abruptly into such notoriety that the Nobel Prize committee pressured her to give up her second award.

Who was the famous "Madame Curie" really, and what did her discoveries mean? Just a few of the big ideas included in this richly entertaining book are:

- How the Curies' work resulted in headlines like "Cure for Cancer Discovered" and "Man-and-Wife Team Discover Perpetual Motion in Hut"
- How Marie Curie discovered the elements radium and polonium
- Why Marie Curie refused to patent her method of producing radium—despite the fact that such a patent would have made her a millionaire
- How the work of the Curies and Henri Becquerel led to the nuclear bomb

Revealing the intensity, dedication and determination of one of the century's most inspiring single working mothers, *Curie and Radioactivity* is for anyone curious about the "female Einstein," and even more important, for all those who want to understand her groundbreaking contributions to nuclear physics—contributions that ultimately cost her her life.